YOUR KNOWLEDGE HAS VALUE

- We will publish your bachelor's and master's thesis, essays and papers
- Your own eBook and book - sold worldwide in all relevant shops
- Earn money with each sale

Upload your text at www.GRIN.com and publish for free

Indigenous Peoples and their role in fighting Climate Crisis in International Climate Politics

Sonja Smolenski

Bibliographic information published by the German National Library:

The German National Library lists this publication in the National Bibliography; detailed bibliographic data are available on the Internet at http://dnb.dnb.de.

ISBN: 9783346550804
This book is also available as an ebook.

© GRIN Publishing GmbH
Nymphenburger Straße 86
80636 München

All rights reserved

Print and binding: Books on Demand GmbH, Norderstedt, Germany
Printed on acid-free paper from responsible sources.

The present work has been carefully prepared. Nevertheless, authors and publishers do not incur liability for the correctness of information, notes, links and advice as well as any printing errors.

GRIN web shop: https://www.grin.com/document/1154522

Faculty of Management, Economics and Social Sciences

Master of Arts Political Science

Term paper subject:

International Climate Politics, Indigenous peoples and their role in fighting climate crisis

Summer term 2021

Submitted by

Sonja Smolenski

Köln, 29th of August 2021

Table of Contents

1. Introduction

Indigenous peoples, who comprise an estimated 400 million people worldwide or 5% of the total population in more than 90 Countries, are one of the communities most affected by climate change and simultaneously play a key role in climate change mitigation through their centuries-old, nature-adapted ways of life. The international community and the climate research community have long overlooked the important role that indigenous peoples have in combating climate change, and have excluded their perspectives from climate negotiations, even though they have the necessary knowledge and manage the most resource-rich areas of the world (Prinz, 2021). 80% of the world's remaining biodiversity is located in indigenous territories, which are in or near 85% of the world's protected areas (Sobrevila, 2008: 5). Indigenous peoples manage at least 17% of the carbon that is stored in global forests. This is equivalent to 33 times the global energy emissions from 2017. Indigenous lands, such as the Amazon basin with its rainforest resources, thus harbor enormous potential to mitigate climate change and its consequences (Schilling-Vacaflor, 2021: 3).

The ways of life of indigenous peoples that have been passed down for centuries could also provide answers to how humans could adapt to climate change in the future and protect natural resources. Indigenous ways of life are closely adapted to their environment – be they in Southeast Asia, the rainforests of Amazonia and Africa, the Arctic, or on the highlands of the Andes. Over centuries, they have developed economic practices that have helped preserve ecosystems and increase biodiversity (Drillisch, 2011: 3).

Although indigenous peoples themselves have contributed least to climate change through their low-carbon lifestyles, they are nonetheless the first to feel its ramifications, according to the fourth report of the Intergovernmental Panel on Climate Change (IPCC) and studies conducted by environmental organizations such as the International Union for Conservation of Nature (IUCN) (Feldt, 2011: 1). They live in ecosystems that are particularly vulnerable to climate change (including humid tropics, polar regions, deserts, small islands, and at high altitudes) as they are heavily reliant upon lands and resources for their livelihoods (for medicine, shelter, and food). Furthermore, they are among the poorest people on the planet, mainly caused by colonialism and ongoing oppressions in the National-States they are living in (Ibid.). As a study by the World Bank shows, climate change increases social inequalities and poverty. This particularly affects indigenous peoples, who make up 15% of the world's poorest population (International Labour Office ILO, 2017: 7-8).

At the first climate conference in 1979 in Geneva, representatives of indigenous peoples warned of the devastating consequences of climate change for their communities and the environment and demanded that their expertise should be included in future discussions, policymaking and negotiations. It took another 30 years until the demands of the indigenous community were heard for the first time by the international community of states at the climate conference in Cancún in 2010, and they were promised a voice in the climate negotiations as so-called "guardians of nature" (Mamo, 2020: 6; Feldt, 2011: 1). In 2021, indigenous organizations are still waiting to be fully included in negotiations at this year's UN Climate Summit, which will take place in Glasgow (Prinz, 2021).

The following paper examines the implementation of indigenous rights in international climate policy to date. First it will show why the climate crisis is a component of colonial continuities in order to better understand what roles indigenous peoples play in international climate policy today. After that, the impact of the climate crisis on the rights of indigenous peoples and why they are at the same time a key figure in the fight against climate change will be revealed. The third section highlights the demands of indigenous associations that are supported by international law. The main section then analyzes previous international climate negotiations and decisions and shows the challenges that have arisen to date in the inclusion of indigenous rights.

2. Indigenous Peoples and Climate Change

2.1 Definitions, research status and postcolonial theory

Although indigenous representatives opposed having an internationally binding definition of indigenous peoples, to preserve the right to self-definition, there is no official definition of indigenous peoples in international criminal law. The former chair of the Working Group on Indigenous Peoples, Erica Daes, points out that "indigenous peoples have suffered from definitions imposed on them by others." Nevertheless, there have been numerous attempts to describe the concept of indigenous peoples. The most frequently cited is from the 1981 study by the Special Rapporteur of the Sub-Commission on Prevention of Discrimination and Protection of Minorities, José R. Martínez Cobo (Office of the United Nations High Commissioner for Human Rights, 2013: 6).

This definition is used for the present paper as it is based on the right of self-definition of indigenous peoples and makes a very broad delimitation:

> "Indigenous communities, peoples and nations are those which, having a historical continuity with pre-invasion and pre-colonial societies that developed on their territories, consider themselves distinct from other sectors of the societies now prevailing on those territories, or parts of them. They form at present non-dominant sectors of society and are determined to preserve, develop and transmit to future generations their ancestral territories, and their ethnic identity, as the basis of their continued existence as peoples, following their cultural patterns, social institutions, and legal system. On an individual basis, an indigenous person belongs to these indigenous populations through self-identification as indigenous (group consciousness) and is recognized and accepted by these populations as one of its members (acceptance by the group). This preserves for these communities the sovereign right and power to decide who belongs to them, without external interference." (Cobo, 1981: 29).

The entanglement of the climate crisis with indigenous rights has long gone unnoticed by mainstream academic discourse. Indigenous associations, as well as postcolonial theories, paved the way for the issue. Indigenous rights were also ignored for a long time in international climate negotiations and treated only as a marginal issue (Kartal et. Al, 2021: 20).

Indigenous scholars and associations hold colonial continuities responsible for this. The postcolonial theory according to Spivak calls the phenomenon epistemic violence. This means that knowledge, in this case Western knowledge, is (forcibly) transferred to non-Western contexts. This is part of colonial continuities, as Chapter 2.2 points out. In this specific case, that means that international climate policy implements Western concepts and governance models while Indigenous knowledge is excluded or even invalidated (Draude, Neuweiler 2010: 1-8).

In climate change policies indigenous knowledge is only used to complement already existing climate change policies with indigenous knowledge. Spivak's criticism is mainly directed against Western experts in the context of development cooperation but also climate research. Indigenous knowledge is said to be included only on the technical level, while the conceptual level is disregarded. Climate protection measures are decided without the involvement of indigenous peoples and their understanding of nature or social justice is not considered. Since international climate policy is part of the governance concept, which is shaped by western governance styles, it is primarily Eurocentric. The following work examines international climate policy from a postcolonial theory according to Spivak (Spivak, 1999: 69).

2.2 Colonialism and Climate Change

"The invading civilization[s] confused ecology with idolatry. Communion with nature was a sin worthy of punishment... Nature was a fierce beast that had to be tamed and punished so that it could work as a machine, placed at our service for ever and ever. Nature, which was eternal, owed us slavery"

(Voskoboynik as cited in Galeano, 2018)

European colonialism can be analyzed on four levels that still occur in colonial continuities today (Voskoboynik, 2018):

- Cultural violence through the imposition of Western concepts (epistemic violence). This was accompanied by the missionization of the Christian religion and the banning of indigenous languages. Today, indigenous languages are threatened with extinction and indigenous culture is considered backward in much of the formerly colonized countries, such as Argentina or Colombia.
- Suppression of local knowledge. Indigenous knowledge was considered regressive and non-scientific. This perception persists to this day. Indigenous peoples have long been seen only as "victims of climate change," while their key role in the fight against climate change and their traditional knowledge for climate research went unnoticed. This is also due to colonial continuities (Fuhr, 2021). Colonialism was legitimized through the invention of racism. The indigenous population was constructed into the "other" and devalued as "uncivilized" in comparison to the white population. Through racist ideology, colonial powers were able to morally legitimize their campaigns. This was accompanied by the devaluation of indigenous knowledge and cultural traditions, which persist to this day and therefore was excluded from scientific discourses and international policy decisions for decades as it was not considered valid, more on this in Chapter 2.4 (Kartal et. Al, 2021: 20-32). In addition, the colonial powers brought with them the idea of perceiving humankind and nature separately, which had to be tamed by people. This subsequently enabled the exploitation of natural resources.
- Economic violence through the exploitation of natural resources of colonized countries. The enslavement and genocide of indigenous peoples in North and South America, Asia and Africa was focusing on the exploitation of natural resources to strengthen the

prosperity of the European colonial powers. During colonization, it was raw materials such as coffee and sugar that were exploited in the colonized areas and brought to Europe. Europe's prosperity, financed by colonialism, later made the Industrial Revolution possible. This is considered the main initiator for the increasing CO2 emissions responsible for the greenhouse effect and the climate crisis (Fuhr, 2021). Today, it is mainly natural mineral resources, such as cobalt and lithium for electrical appliances, that are exploited. These are partly located on the territories of the remaining indigenous population and lead to a massive exploitation and displacement of the local indigenous population and at the same time to the aggravation of the climate crisis by destroying habitats and the environment (Fuhr, 2021).

- political violence through the repression of the local population (Voskoboynik, 2018).

The climate crisis, international climate policy and indigenous rights must be seen in the context of European colonialism and racism. The climate change is often described as man-made or human-driven. However, climate change caused by increasing emissions and greenhouse gases is not equally attributable to the entire global population. Historically, the Global North has been the largest emitter of climate-damaging greenhouse gases compared to the Global South. 19% of the world's population in the Global North is responsible for 92% of global CO2 emissions, with the final 8% distributed among the rest of the world's population (Kartal et. Al, 2021: 20-32). At the same time, countries in the Global South are the most affected by the consequences of climate change, as they are among the poorest countries in the world. This is also a direct consequence of colonialism, which exploited the mineral resources and labor of the affected countries for centuries. Left behind were weak nation-state systems, corrupt governments, and marginalized minorities that cannot be protected enough from climate change (Ibid.).

Colonial Continuities pursue to influence international climate policy. This is particularly evident in United Nations-led climate negotiations and multi and bilateral climate movement organizations, which are disproportionately staffed by industrialized countries or former colonial powers. International climate protection projects are mostly implemented without the consent of local, indigenous populations, more on this in Chapter 4.2 (Fuhr, 2021). It is crucial to include the perspective of colonial continuities in order to sustain the climate crisis and indigenous rights inclusion at the international level.

2.3 Effects of Climate Change on Indigenous Peoples rights

> *"Indeed, climate change poses a direct threat to a wide range of universally recognized human rights, such as the rights to life, food, adequate housing, or water. Procedural human rights, including access to information or justice and participation in decision-making processes, may also become increasingly relevant in a context of climate change, particularly for those being affected by it."* (Ms. Kyung-wha Kang, Deputy of the United Nations High Commissioner for Human Rights in her speech to COP 13, Bali, December 2007).

Any change in the climate has a devastating impact on the lives of indigenous communities. As a result, their sovereignty and fundamental rights are threatened. In her report addressed to the Commission on Human Rights, Françoise Hampson, former United Nations expert to the Sub-Commission on the Promotion and Protection of Human Rights, outlines three levels at which indigenous peoples' rights are uniquely affected by climate change. These include (Feldt, 2011:5):

- Environmental damage and changes in climate affect life in their territories, due to factors such as water scarcity or loss of biological diversity. Hampson assumes that indigenous peoples can only exercise their individual and collective rights to a limited extent (Feldt, 2011: 5). These rights include their food security, which threatens their sovereignty. In the Arctic regions, for example, indigenous peoples rely upon fishing as well as hunting seals, walruses, and polar bears, practices that they have perfected over centuries in harmony with the environment. The increasing melting of the polar ice caps and rising sea levels are likewise destroying animals' habitats that are now threatened with extinction due to climate change. Indigenous communities are thus losing their main source of food (The United Nations Department of Economic and Social Affairs, n.d.).
- Being displaced from their land is also an issue, which is due to land grabbing or flight resulting from climate change to other regions of the same state. Deforestation and land exploitation cause the displacement and the destruction of indigenous people's habitat. Many indigenous communities are displaced within their own lands. They are considered internally displaced within their own countries and often must flee to the peripheries of large cities, where they are forced to live in precarious conditions. As a result, they lose their traditional

housing and protection rights, as they can no longer live on their territory (Feld, 2011: 5).

- The third issue is expulsion from their land and forced expulsion to another state. It is to be expected that they will lose all of their rights as indigenous peoples as a result, as they are no longer considered as indigenous but as an ethnic minority in a foreign country by ILO Convention 169. They thus lose all of their rights as indigenous peoples, such as the right to their territory. This particularly affects the population of the island states in the Pacific, whose habitat is acutely threatened by flooding (Ibid.).

Another human right that is under significant pressure is the right to life. Environmental crimes particularly affect indigenous communities, whose resource-rich territories have become a target for armed actors. Indigenous peoples who protect their territories often pay with their lives. In 2020, a total of 220 environmental activists were murdered globally (Global Witness, 2020: 17). In Colombia, one indigenous person is murdered every 72 hours for trying to protect their territory from land grabbing (Díaz, 2019).

In Brazil and Paraguay, the Guarani indigenous people have defended their land against soy farmers and ranchers who are destroying their habitat for pasture and crops without their consent. The Guarani have called for a boycott of internationally exported beef and soy products. Since then, countless members of their community have been murdered. In 2015, two youths were shot dead by farmers (Survival, 2015).

Climate change thereby exacerbates the precarious living conditions of indigenous peoples, who are already impacted by human rights violations, discrimination, and political and economic marginalization in nation-states (The United Nations Department of Economic and Social Affairs, n.d.).

These examples show that international climate protection and climate policy must be considered in connection with the protection of human rights in order to be able to sufficiently protect both indigenous communities and the world's remaining biodiversity (Prinz, 2021).

2.4 Combating climate change with the help of indigenous knowledge

"Indigenous, local, and traditional forms of knowledge are a major resource for adapting to climate change. Natural resource dependent communities, including indigenous peoples, have a long history of adapting to highly variable and changing social and ecological conditions. But the salience of indigenous, local, and traditional

knowledge will be challenged by climate change impacts. Such forms of knowledge are often neglected in policy and research, and their mutual recognition and integration with scientific knowledge will increase the effectiveness of adaptation."

(Adger et. Al. 2014: 758).

The term "indigenous knowledge" or traditional knowledge refers to the knowledge that indigenous communities have cultivated over centuries and passed on through the majority oral tradition within the communities. It stands as an antithesis to scientific knowledge, which is also referred to as "Western knowledge." Indigenous knowledge has long been excluded from mainstream research and considered a fringe discipline of science due to the degradation caused by colonialism (Bruchac, 2014: 1).

Since indigenous knowledge has so far only been included to complement Western science, the following examples are problematic from a postcolonial perspective in that they include indigenous knowledge systems only at the technical level and exploit them for climate research. This reproduces existing power in a postcolonial context, as knowledge hierarchies emerge. While climate research only uses concepts from indigenous knowledge that are valid from their perspective, other traditional knowledge systems are disregarded. A mere addition of indigenous knowledge systems to already existing research, disregards the conflicts that can arise when different concepts of development and environment clash. A condition for equal exchange between Western and indigenous knowledge systems would be that Climate Science recognizes indigenous knowledge as equal (Draude, Neuweiler 2010: 8).

The knowledge of indigenous peoples is fundamental stopping climate change. Indigenous lifestyles are sustainable and, in contrast to Western industrialized lifestyles, hardly consume any carbon, as shown, for example, in the management of tropical rainforests by indigenous communities or traditional livestock farming systems (Feldt, 2011: 5).

Indigenous economies depend primarily on the ecosystems in which they live and on natural resources that they process sustainably. In addition, their cultural and traditional ways of life are based on a close connection with nature. Thus, they contribute significantly to the protection of ecosystems (ILO, 2017: 23).

According to their tradition, for example, it is strictly forbidden for the indigenous population of the Qashqai in Iran to cut down trees. The economic base of the Qashqai are non-timber products such as rubber and medicinal plants. The sustainable hunting practices of the Qashqai have also helped to sustain wildlife populations in the region. The nomadic peoples have contributed to climate adaptation strategies by storing water in dry areas for their livestock and have developed early warning systems that allow them to predict droughts and take preventative measures. Their knowledge plays a fundamental role in climate change adaptation and extreme weather forecasting, which are of great importance to science (Ibid.).

Climate-just agriculture, which is used by the Food and Agriculture Organization of the United Nations in the fight against climate change, is based on indigenous knowledge that has been transferred to the databases with the help of indigenous associations. One example is the cultivation method of indigenous communities in the Mekong Delta in Vietnam, who sow a wild rice variety on land that is frequently affected by flooding. The method protects the land from increasing flooding (Ibid.: 27).

In addition, research shows that indigenous peoples can adapt to extreme climate changes through their traditional knowledge and predict climate disasters through their accurate weather forecasts. In 2004 the indigenous Moken peoples of Myanmar were the first ones to warn about the tsunami happening in the Indian Ocean after observing how the water receded on the shores. This saved the entire village from the environmental disaster. In addition, their houses, which are built of traditional materials such as bamboo and thatch, are not lethal to the residents in the event of a collapse. The Simeulue community in Indonesia was also able to predict the tsunami through their traditional knowledge, saving more than 80,000 people. They were awarded the Sasakwa Prize by the United Nations. Today, many early warning systems for tsunamis rely on the knowledge of indigenous peoples (Ibid.)

These are just a few examples that show how relevant indigenous knowledge is in the fight against climate change. Many climate researchers have now recognized the contribution and are working together with indigenous communities. Only on the political level, the cooperation could not be equally enforced so far, which is due to the political marginalization of indigenous peoples (Feldt, 2011: 4).

3. Indigenous Peoples and International Climate Politics

3.1 Demands of Indigenous Peoples in Climate Negotiations

Indigenous communities around the world have formed various associations to make common demands of the international community. One of the primary demands is that they are involved in climate projects of the international community, and that they be able to exercise their prior, voluntary, self-determined, and informed consent on planned climate projects that affect their indigenous territories. This means that indigenous communities can, for example, reject projects before they occur (Feldt, 2011: 1).

A second key demand is that they be able to participate and be included in all climate change plans and programs. To date, indigenous communities have been largely excluded from measures that affect their habitat. The present demand is that indigenous peoples be included in processes in the future, such as the implementation and monitoring of national and international climate science research. One example is the Reducing Emissions from Deforestation and Degradation (REDD) program that was adopted at the 2007 Conference of the Parties in Bali. Given that between 16% to 20% of global CO2 emissions are caused by the destruction of forests, the program is designed to help support forest conservation via policy and funding measures. The program directly affects indigenous communities, because indigenous territories are threatened by increasing deforestation, such as in the Amazon rainforest. At the same time, they have important knowledge regarding how to preserve forests. Indigenous associations are therefore calling for participation in the implementation of such programs like REDD (Drillisch, 2011:6).

Another demand of global indigenous associations is the recognition and inclusion of their traditional knowledge in climate negotiations. Indigenous representatives have only been involved in climate research for a relatively brief amount of time, though their traditional knowledge has been cultivated over generations and could make an important contribution to climate protection, as Chapter 2.4 revealed. Even though indigenous peoples emphasize that they want to contribute their knowledge to climate research, they simultaneously demand that their intellectual property rights be protected (Ibid.: 7).

The protection of indigenous territories is another relevant demand. Indigenous territories are threatened by land grabbing and the increasing destruction of their habitat, as outlined in Chapter 2.1. A key demand for climate regulation is that the international community of states, or the nation-states in which indigenous territories are located, ensure that communities are protected. This is fundamental not only to protect indigenous communities' human rights, but also the environment that is being destroyed for profit in the process of land grabbing and dispossession. To date, indigenous communities protect their territory chiefly in a self-organized way (Ibid.).

Another central demand is that the international community of states reach an agreement that enforces a binding obligation to reduce CO2 emissions by 95% by 2050. These figures are in line with the demands of the environmental protection associations and the IPCC calculations from 1990 (Ibid.).

Furthermore, indigenous people demand that the territories of indigenous communities be legally recognized and protected as a separate category of protection. Their territories should fall under the protection category of "Strict Nature Reserve" (Ibid.) A study from the United Kingdom that examined the effectiveness of protected areas for the resilience to harmful human impacts found that not every type of protected area protects biodiversity. Out of 12,000 protected areas in more than 152 countries, the study found that what is most needed for effective conservation is the consolidation of human rights in protected areas. The people living in such areas protect the territories from illegal interference undertaken by companies or individuals. The study revealed that the control by indigenous peoples was much more effective than by the nation-states themselves (Prinz, 2021).

3.2 Legal Framework

The demands of indigenous peoples are based upon several internationally signed documents that are binding under international law. These include the UN Covenant on Economic, Social and Cultural Rights signed in 1966, which recognizes the right to self-determination of peoples and states that no person shall be deprived of their means of subsistence; ILO Convention 169 for Indigenous and Tribal Peoples, 1989; and the United Nations Declarations of the Rights of Indigenous Peoples (UNDRIP), adopted by the United Nations General Assembly in 2007. Article 19 of the United Nations Declaration recognizes the right to free, early, and informed consent in all projects that affect indigenous territory, as well as adequate compensation if violated. The 1989 and 2007 declarations grant indigenous peoples the right to own land in Article 26, and the right to self-determination in Articles 3 and 4 (Feldt, 2011: 5).

The UN Convention on Biological Diversity (CBD) outlines the special importance of indigenous peoples and their traditional knowledge for the sustainable use of biological resources and the preservation of the environment, more generally. However, the recognition of the rights of indigenous peoples has been lacking in climate framework conventions to date.

In an attempt to change this fact, indigenous associations founded the International Peoples Forum on Climate Change in 2000 in order to be able to influence the negotiations of the signatory states and to advocate for their rights (Drillisch, 2011: 5-6).

A chance to provide indigenous peoples a greater say was made by the UN in 2017. More specifically, the status of indigenous peoples at climate conferences was equated with that of non-governmental organizations (Etchart, 2017: 3).

However, this is not sufficient to fully include indigenous peoples in climate negotiations. The primary challenge in recognizing their rights from a legal perspective in this case is the structure of international law. According to existing international law, indigenous peoples, much like economic interest groups and private organizations, do not have the same status as states because they are not subjects of international law. This means that they are only granted an advisory function but no decision-making function in climate convention negotiations and are thus excluded from a majority of decisions (Prinz, 2021). Even though the international community of states refers to the rights of indigenous peoples and their knowledge in the final document from the 16th Conference of the Parties to the Framework Convention on Climate Change in Cancuín in 2010, the anchoring of their rights has to date been lacking in the context of international climate negotiations. Even at the level of nation-states, individual countries are required to implement a real anchoring of indigenous rights in action plans for climate protection (Feldt, 2011: 7).

4. Implementation and Challenges of Indigenous Rights in Climate Conference Negotiations and Measures

4.1 Implementations of Indigenous Rights in Climate Conference Negotiations

International climate negotiations have long disregarded the contribution of indigenous peoples to combating climate change, even though international documents from both the UN and IPCC have repeatedly affirmed the fundamental relevance of indigenous actors to climate change (Sherpa, 2019).

When the heads of state and government of 195 countries agreed on a legally binding agreement in Paris in December 2015 following two decades of climate negotiations in the framework of the UN Framework Convention on Climate Change, over 250 indigenous delegates were also present. They had lobbied for the rights of indigenous peoples to be enshrined in the agreement.

The Paris Agreement was the first document to ever include the rights of indigenous peoples without qualification in a legally binding UN treaty (Cultural Survival, 2015). However, what was considered as one of the most important legally binding climate agreements nonetheless disappointed indigenous representatives, because it did not include any legally binding references to the protection of indigenous peoples' rights and sovereignty. Overall, only nonbinding references to indigenous populations were written into the Paris Climate Agreement: preamble paragraph 8, preamble 16, page 19 Non-treaty Parties paragraph 136, as well as preamble 12 page 20 and article 7 paragraph 5 page 24) (Ibid.)

It was four years later when a historic outcome for indigenous people was actually achieved at the UN Climate Change Conference 2019 in Madrid when the work plan of the Facilitative Working Group of the Local Communities and Indigenous Peoples Platform (LCIPP) was adopted. The working group stems from the International Indigenous Peoples' Forum on Climate Change (IIPFCC), established in 2008, which is the formal framework for the Indigenous Peoples' Negotiating Body at the UN Climate Conferences. The working group is comprised of a representative from a small island state, a representative of a party from a least developed country, seven representatives of indigenous organizations, and representatives from the regional groups of the United Nations (Sherpa, 2019). The following demands are to be implemented by the working group as part of the climate agreements starting in January 2020: the inclusion of indigenous knowledge in climate negotiations and the strengthening of the participation of indigenous associations in political action relevant to policy and the climate crisis (Ibid.). The meeting was historic in the sense that it was the first time ever that representatives of indigenous peoples were given the same space as state actors in international climate negotiations (Ibid.)

4.2 Challenges of the Implementations of Indigenous Rights in Climate Conference Negotiations and Measures

UN climate conferences have to date focused on the energy policies of industrialized nations rather than on the protection of natural habitats, such as the Amazon rainforest. Indigenous peoples received no support from the UN climate conference prior to the Paris Agreement. A report by the Rights and Resources Initiative reveals that, out of a total of 47 countries where indigenous people live, 26 countries make no reference to lands managed by indigenous communities in their climate change proposals during the UN climate conference in Paris (Survival International, 2015).

The Paris Agreement has been widely criticized because earlier drafts initially included the protection of indigenous peoples' rights in Article 2.2. Due to instructions of the United States, the European Union and Norway, the article was removed from the legally binding section, even though these states have a leading role in the fight against climate change. Only following pressure from concerned organizations was a reference to the protection of indigenous peoples added to the preamble of the Paris Agreement (Cultural Survival, 2015). The UN Special

Rapporteur on the Rights of Indigenous Peoples, Victoria Tauli-Corpuz, welcomed the decision, but criticized the fact that the reference to the protection of indigenous rights was quite weak in its wording. The semantics of the text indicate a further dilution of indigenous peoples' rights. For example, all of the obligations that follow the verb "shall" are legally binding, but those that follow "should" are not. Only the phrase "should" is used in the preamble. For this reason, the protection of indigenous peoples in the Paris Agreement is not legally binding but rather constitutes a recommendation for action by nation-states (Ibid.).

Climate measures adopted in the course of climate negotiations also do not include indigenous rights, at least to date. Internationally agreed upon climate protection measures can – like climate change – have a serious impact on the living conditions of indigenous peoples. So far, the sustainable and customary use by indigenous peoples is not protected and, in some countries, certain practices are even classified as illegal. (Prinz, 2021).

One example that shows that the enforcement of international climate protection regulations often takes place without the inclusion of indigenous realities is the High Ambition Coalition for Nature and People (HAC) initiative that was launched as part of the Global Deal for Nature. It was adopted in January 2021 by over 50 countries, including Germany. It aims to protect 30% of the total marine and terrestrial areas that are of particular importance for biodiversity by 2030. The target is also mentioned in the UN Convention on Biological Diversity under point two. The current draft law does not include any reference or safeguard clause for the recognition of the rights of indigenous people who are actually responsible for the conservation and maintenance of these territories (Ibid.). The draft law poses a risk to the rights of indigenous peoples, as it may lead to the expropriation of indigenous lands in the name of climate change mitigation and the eviction of communities from their territories so that they can be converted into protected areas. This has already occurred in the past, as an open letter from the human rights organization Survival International to German Chancellor Angela Merkel shows. This form of conservation has led to the expulsion of indigenous peoples from their territories in both Asia and Africa. In an interview, one of the initiators explains that the notion of establishing nature reserves goes hand in hand with the idea of 'untouched nature', which means land without humans, that is part of a postcolonial perspective. The global expansion of the concept of the nature reserve also goes hand in hand with the practice of criminalizing the living practices of indigenous peoples and driving them out of their homelands (Ibid.).

Internationally adopted forest conservation projects, such as the REDD program, also threaten to violate Indigenous Peoples' rights by, among other things, restraining access to land and food.

The REDD program is a mechanism adopted as part of the UN climate negotiations. The program aims to improve forest carbon stocks in global south countries and reduce emissions from deforestation and forest degradation. At the same time, the program aims to guarantee the conservation and sustainable management of forests (Schielmann, S. et. Al, 2013: 8). As REDD creates incentives to make money from forests, indigenous associations fear that nation-state governments will be more interested in forest areas rather than indigenous rights and compliance. This could, for example, result in new protected areas like carbon reservoirs without consideration of the other functions of forests. Traditional economies, such as firewood gathering, could thus become criminalized to a greater extent. There is a threat of significant land grabbing and expropriation, secured on a legal basis through conservation agreements, that further marginalizes indigenous peoples (Feldt, 2009: 6).

Incorporating a human rights approach to all forest conservation or climate change mitigation projects would mean co-creating every project with local forest people. This includes addressing their customary rights to their traditional lands, resources, and cultural beliefs. The approach is grounded in international human rights standards based on the United Nations Declaration on the Rights of Indigenous Peoples. Its primary purpose is to ensure that the rights of indigenous forest peoples be respected, following the belief that indigenous peoples play a key role in the management and protection of forests. To date, most climate change mitigation projects have failed to implement indigenous collective rights (the right to self-determination and political participation, rights of disposition over jointly managed lands and natural resources), and have thus disregarded the protection of indigenous rights in conservation efforts (Schielmann et. Al, 2013: 23).

5. Conclusion

> "*The failure of the United Nations to protect the rights of Indigenous Peoples in the final agreement will further fuel the destruction of forests and other ecosystems that have always been managed by Indigenous Peoples.*" (Prince, 2021)
>
> UN Rapporteur on Indigenous Rights Victoria Tauli-Corpuz

The following work shows that indigenous peoples are environmentalists and guardians of nature. At the same time, they are most acutely threatened by the climate crisis. Without the support of the international community and inclusion in international climate policy, their habitat, and thus the largest deposits of biodiversity in the world, will soon be extinct (Survival, 2015).

The consequences of the exclusion of indigenous peoples in policymaking and climate conferences is that global projects aiming to mitigate the impacts of climate change are undermining the livelihoods of indigenous peoples and their realities (Raygorodetsky, 2011).

In order to keep the promises made by the international community regarding the reduction of emissions by 2030, only implementing protective measures for the environment is not enough. The situation of indigenous peoples, who stand as both the most affected and key players in the fight against climate change, highlights the need for international climate policy to go hand in hand with human rights, while at the same time taking place from a decolonial perspective (Survival, 2015). Specifically, this means that vulnerable populations, their habitats, and their traditions must be prioritized in the design and implementation of climate agreements, and their knowledge must be equated with Western science (Spivak, 1999: 73).

While the climate conference is taking place in Glasgow in November 2021, various non-governmental organizations and indigenous associations are meeting in Marseille in September for the first Congress for the Decolonization of Conservation. They call on the international community to stop previous conservation programs. One point of criticism is for example, the demand to convert 30% of the earth's surface into protected areas. That act, in their opinion, would not stop the loss of biodiversity or the climate crisis but instead lead tot the oppression of the local, indigenous population and tie in with (the idea of) colonialism. They advocate for the equal inclusion of indigenous people, whom they call the best conservationists in the world, in international climate policy (Our Land, Our Nature 2021). It is hoped that the congress will be able to influence the negotiations of the international community and the future of international climate policy.

Bibliography:

Adger W.N., Pulhin J.M., Barnett J. et al.: "Human security", in C.B. Field, V.R. Barros, D.J. Dokken et al. (eds): Climate Change 2014: *Impacts, Adaptation, and Vulnerability. Part A: Global and Sectoral Aspects.* Contribution of Working Group II to the Fifth Assessment Report of the Intergovernmental Panel on Climate Change (Cambridge and New York, Cambridge University Press, 2014), pp. 775–791.

Bruchac, Margaret (2014): *Indigenous Knowledge and Traditional Knowledge.* In Encyclopedia of Global Archaeology. Claire Smith, ed., chapter 10, pp. 3814-3824. New York, NY: Springer Science and Business Media.

Cobo Martínez R., José (1981): *Study of the Problem of Discrimination against Indigenous Population.* Volume V. Conclusions, Proposals and Recommendation. United Nations, New York. https://cendoc.docip.org/collect/cendocdo/index/assoc/HASH01a2/55590d02.dir/Martinez-Cobo-a-1.pdf

Cultural Survival (2015): *"Annexed:" The Rights of Indigenous Peoples in the UN Climate Change Conference 2015.* www.culturalsurvival.org/news/annexed-rights-indigenous-peoples-un-climate-change-conference-2015 [Visited: 29.07.2021]

Díaz González, Marcos (2019): *Asesinatos de indígenas en Colombia: "Es un genocidio", 6 claves para entender los crímenes en el Cauca* https://www.bbc.com/mundo/noticias-america-latina-50341874 [Visited: 26.07.2021]

Draude, Anke& Neuweiler, Sonja 2010: Governance in der postkolonialen Kritik. Die Herausforderung lokaler Vielfalt jenseits der westlichen Welt, SFB-Governance Working Paper Series, Nr. 24, DFG Sonderforschungsbereich 700, Berlin, Mai 2010. ISSN 1864-1024 (Internet)

Drillisch, Heike (2011): *Indigene Völker und die Klimaverhandlungen. Die Beschlüsse der Klimaverhandlungen in Kopenhagen und Cancún: ihre Bedeutung für indigene Völker und Anknüpfungspunkte für Umweltorganisationen.* (INFOE – Institut für Ökologie und Aktions-Ethnologie e.V., Ed.). https://www.infoe.de/images/Pdf/INFOE_Indigene_%20Volker_Klimaverhandlungen.pdf

Etchart, Linda (2017): *The role of indigenous peoples in combating climate change.* www.researchgate.net/publication/319234804_The_role_of_indigenous_peoples_in_combating_climate_change

Feldt, Heidi (2009): *Indigene Völker und Klimapolitik. Überblick über die aktuelle Situation und indigene Positionen im Kontext der Verhandlungen zur Klimarahmenkonventionen.* Deutsche Gesellschaft für Technische Zusammenarbeit (gtz).

Feldt, Heidi Dr. (2011): *Stärkung Indigener Organisationen in Lateinamerika: Indigene Völker und Klimawandel. Zusammenhang zwischen Klimawandel und indigenen Völkern und ihre Position im Kontext der Verhandlungen zur Klimarahmenkonvention.* Deutsche Gesellschaft für Internationale Zusammenarbeit (Editor).

Fuhr, Lili (2021): *Wie Rassismus und Kolonialismus die Klimakrise und den Klimaschutz prägen. https//:klima-der-gerechtigkeit.de/2021/03/19/wie-rassismus-und-klimawandel-miteinander-verwoben-sind/* [Visited: 13.08.2021]

International Labour Office, ILO (2017): *Indigenous peoples and climate change. From victims to change agents through decent work.* Geneva. www.ilo.org/wcmsp5/groups/public/---dgreports/---gender/documents/publication/wcms_551189.pdf

Galeano, Eduardo (n. Y): *Mundo: Cuatro frases que hacen crecer la nariz de Pinocho.* www.servindi.org/actualidad/20810 [Visited: 11.08.2021]

Kartal, Shayli et. Al. (2021*): Kolonialismus und Klimakrise. Über 500 Jahre Widerstand.* Jugend im Bund für Umwelt und Naturschutz Deutschland e.V (Editor). Berlin. www.bundjugend.de/wp-content/uploads/Kolonialismus-und-Klimakrise-Ueber-500-Jahre-Widerstand-11.pdf

Kyung-wha, Kang (2007): "Climate Change and Human Rights". Office of the High Commissioner for Human Rights. Conference of the Parties to the United Nations Framework Convention on Climate Change and its Kyoto Protocol. 3-14 December 2007. Bali, Indonesia. https://newsarchive.ohchr.org/EN/NewsEvents/Pages/DisplayNews.aspx?NewsID=200&LangID=E [Visited: 01.08.2021]

Mamo, Dwayne (2020): The indigenous World 2020, 34th Edition. The International Work Group for Indigenous Affairs (IWGIA), 2020.

Office of the United Nations High Commissioner for Human Rights. Indigenous Peoples and Minorities Section (2013): *The United Nations Declaration on the Rights of Indigenous Peoples. A Manual for National Human Rights Institutions.* Asia Pacific Forum of National Human Rights Institutions and the Office of the United Nations High Commissioner for Human Rights August 2013 https://www.ohchr.org/documents/issues/ipeoples/undripmanualfornhris.pdf

Our Land, our Nature (2021): *A Congress to decolonize Conservation.* www.ourlandournature.org [Visited: 05.08.2021]

Prinz, Ulrike (2021): *Umweltpolitik: Indigene fordern wirksamen Schutz ihrer Rechte, Lebensweisen und Territorien.* https://www.riffreporter.de/de/umwelt/indigene-fordern-bei-naturschutz-klima-umwelt-rechte-mitbestimmung [Visited: 02.08.2021]

Raygorodetsky, Gleb (2011): *Why Traditional Knowledge Holds the Key to Climate Change.* United Nations University. unu.edu/publications/articles/why-traditional-knowledge-holds-the-key-to-climate-change.html [Visited: 16.08.2021]

Schielmann, S., Degawan, M., Falley-Rothkopf, E., Henneberger, B., Mantzel, K. & Nolte, U. (2013): *Waldschutzvorhaben im Rahmen der Klimapolitik und die Rechte indigener Völker* (INFOE – Institut für Ökologie und Aktions-Ethnologie e.V, Ed.). https://www.infoe.de/images/stories/pdf/infoe_waldstudie_final_net.pdf

Schilling-Vacaflor Dr., Almut (2021): *Indigene Völker und Umweltschutz.* Deutscher Bundestag. Parlamentarischer Beirat für nachhaltige Entwicklung. Ausschlussdrucksache 19(26)125. Universität Osnabrück. Nachhaltigkeitsbeirat. https://www.bundestag.de/resource/blob/846220/a33590deb63d523322db4186d6b19399/PowerPoint-Praesentation-von-Frau-Dr-Almut-Schilling-Vacaflor-data.pdf

Sherpa, Pasang Dolma (2019): *The Historical Journey of Indigenous Peoples in Climate Change Negotiation. www.iucn.org/news/commission-environmental-economic-and-social-policy/201912/historical-journey-indigenous-peoples-climate-change-negotiation* [Visited: 21.08.2021]

Sobrevila, Claudia (2008): *The Role of Indigenous Peoples in Biodiversity Conservation. The Natural But Often Forgotten Partners*, The World Bank, New York.

Spivak, Gayatri Chakravorty 1988/1994: Can the Subaltern Speak? in: Williams, Patrick/Chrisman, Laura (Hrsg.): Colonial Discourse and Postcolonial Theory. A Reader, New York, NY, 66- 111

Survival International (2015): *COP 21: Indigene kämpfen allen voran gegen den Klimawandel.* *www.survivalinternational.de/nachrichten/11018* [Visited: 02.08.2021]

The United Nations Department of Economic and Social Affairs. Department of Economic and Social Affairs. Indigenous Peoples (no Date): *Climate Change.* www.un.org/development/desa/indigenouspeoples/climate-change.html [Visited: 11.08.2021]

UN Human Rights Council (2009): Resolution 10/4. *Human rights and climate change*. Tenth Session. https//:ap.ohchr.org/documents/E/HRC/resolutions/A_HRC_RES_10_4.pdf

Voskoboynik, Daniel Macmillen (2018): *To fix the climate crisis, we must face up to our imperial past.* www.opendemocracy.net/en/opendemocracyuk/to-fix-climate-crisis-we-must-acknowledge-our-imperial-past/ [Visited: 23.08.2021]

YOUR KNOWLEDGE HAS VALUE

- We will publish your bachelor's and master's thesis, essays and papers
- Your own eBook and book - sold worldwide in all relevant shops
- Earn money with each sale

Upload your text at www.GRIN.com and publish for free